IMAGES
of America

Missouri Wine Country

St. Charles to Hermann

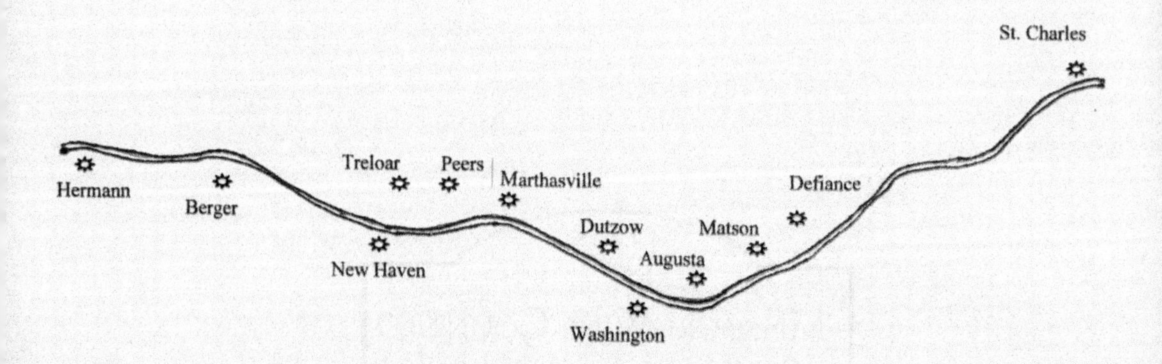

A hand-drawn map of the Missouri River area from St. Charles to Hermann shows the geographic relationship of each of the towns along the way. The Missouri Weinstrasse is the stretch of Missouri wine country north of the river from Defiance to Dutzow (some say Marthasville), and the Herman Wine Trail is the trail of wineries south of the Missouri River from New Haven to Hermann. (Authors' collection.)

On the Cover: Jim and Betty Held, owners of Stone Hill Winery in Hermann, Missouri, enjoy the fruits of their vineyards during the late 1960s. The Hungry Five, a Hermann band, accompanies the wine tasting. (Courtesy of Historic Hermann, Inc.'s Museum at the German School.)

IMAGES
of America

MISSOURI WINE COUNTRY
ST. CHARLES TO HERMANN

Dianna Graveman and Don Graveman

ISBN 978-1-5316-5141-1

Published by Arcadia Publishing
Charleston, South Carolina

Library of Congress Control Number: 2009943866

For all general information contact Arcadia Publishing at:
Telephone 843-853-2070
Fax 843-853-0044
E-mail sales@arcadiapublishing.com
For customer service and orders:
Toll-Free 1-888-313-2665

Visit us on the Internet at www.arcadiapublishing.com

For Dianna's father, Bruce Musterman, who supports us in everything we do, and in memory of Dianna's mother, Doris Musterman, whose spirit will be with us always.

Contents

ACKNOWLEDGMENTS

For advice and support, many thanks to Todd Christine, Harold and Gina Christine, Don and Aggie Graveman, Bruce and Doris Musterman, Vicki Erwin at Main Street Books, Anita Mallinckrodt, and our many friends at Saturday Writers and St. Louis Writers Guild. Thanks to our kids, Steve, Beth, and Teresa Graveman, for cheerfully putting up with their parents' ongoing quest to be writer-historians.

For taking the time to educate us and spending many hours helping us find photographs in their archives and filling in the missing details, we owe thanks to Horace Hesse and Lois Puchta at Gasconade County Historical Society; Carol Kallmeyer, Donna Layman, Jon Layman, and Phyllis Robinson at Historic Hermann, Inc.'s Museum at the German School; Alan Bell at New Haven Preservation Society; Cleta Flynn and Bill Popp at St. Charles County Historical Society; Betty Held, Jim Held, and Lucinda Huskey at Stone Hill Winery in Hermann; Marie Hollenbeck and Terri Werges at Warren County Historical Society; and Bridgette Epple, George Bocklage, and Marc Houseman at Washington Historical Society.

INTRODUCTION

Missouri wine country owes its beginnings to Gottfried Duden, a German researcher and explorer who settled along the Missouri River in 1824 near what is now Dutzow. His famous book, composed as a series of letters, was titled, *Bericht über eine Reise nach den westlichen Staaten Nordamerika's* (Report of a Journey to the Western States of North America). In it, Duden glowingly described the Missouri River Valley between St. Louis and Hermann, and compared the Missouri River to the Rhine in Germany.

Gottfried Duden returned to Germany in 1827, and his book was published in 1829. The book influenced many of Duden's countrymen to begin immigrating to Missouri in the 1830s, and by 1860, more than 38,000 Germans had settled in the lower Missouri River Valley. These immigrants to Missouri are sometimes referred to as "followers of Duden."

About the same time, in 1837, the German Settlement Society of Philadelphia founded Hermann. The society's members hoped to establish a colony where their German language and customs could be preserved. The town was named for Hermann der Cherusker, a German folk hero who led a successful battle against the Romans in 9 AD that changed the course of history. By the mid 1840s, grape-growing experiments had begun in Hermann. The first crop succeeded in 1845, and in 1846, the first wine was made. Stone Hill Winery, still in existence today, was established in Hermann in 1847 by Michael Poeschel. The town celebrated its first Weinfest with an elaborate parade in the fall of 1848.

In 1859, George Münch, another German immigrant, moved from Dutzow to Augusta and began to grow grapes. By 1867, the Augusta grape-growing business was doing so well that the first cooperative wine-making effort in the Missouri River Valley was organized and became the Augusta Wine Company. Mount Pleasant Winery was founded in 1889, and the Missouri wine country boom was well underway.

Prohibition, which forbade the sale and consumption of alcoholic beverages during the 1920s, threatened to destroy the wine-making industry. In fact, wine-making halted in Missouri for almost half a century. However, a man from Illinois named Lucian Dressel eventually bought the Mount Pleasant Winery property in the 1960s. Together with Clay Byers, who owned Montelle Vineyards, Dressel was instrumental in getting Augusta named America's first wine district in 1980. Restoration had also begun on the old winery properties in Hermann during the 1960s, and the Hermann American Viticultural Area was officially designated in 1983. Missouri wine country was once again on its way up.

Today the area is known as home to the famous Missouri Weinstrasse, a two-lane "wine road" that winds through the woods and valleys of southeast St. Charles County and is home to several wineries and vineyards. On the south side of the Missouri River is the Hermann Wine Trail, which stretches 20 miles along the river between Hermann and New Haven and includes several family-owned vineyards and wineries.

However, the fascinating history of this area isn't limited to the account of wineries and winemaking. Most of the towns along the Missouri River from St. Charles to Hermann were settled by German immigrants during the 1800s, and each has a rich, unique story of its own.

St. Charles (originally named *Les Petites Côtes* or "The Little Hills") was settled in 1769 by French-Canadian fur trader, Louis Blanchette. Lewis and Clark began their famous expedition here in 1804, and the city served as Missouri's first state capital from 1821 to 1826. The settlement was later renamed for San Carlos Borromeo, the patron saint of Spain's reigning king. The city was incorporated in 1809.

Defiance was initially going to be called Parsons for a landowner. However, when the Katy (Missouri-Kansas-Texas) Railroad arrived in the 1800s, there was another town named Parsons on the Katy line in Kansas. Alternate names were considered, including Missouriton and Bluff City. The town's settlers eventually decided to name their new home Defiance in honor of their rivalry with Matson, Missouri, to get a station on the line.

The town of Augusta was founded in 1836 by Leonard Harold, one of the settlers who followed Daniel Boone to Saint Charles County. Harold laid out the town of Mount Pleasant, as it was originally known, on part of the government land he purchased in 1821. The town was incorporated in 1855.

Dutzow is the oldest German settlement in the state. It was founded in 1832 by Johann Wilhelm Bock, who named the town for his estate in Mecklenberg, Germany. Bock's Berlin Society members were some of the first followers of Gottfried Duden. Another famous resident was Friedrich Münch, George's brother, who led the Giessen Society in 1834.

The original grave of pioneer Daniel Boone is located in Marthasville. In 1845, Boone's remains were supposedly moved to Kentucky for burial, but some believe the wrong body or not all of Boone's body was moved and that at least some of the pioneer's remains are still buried in Missouri. Marthasville is considered by many to mark one end of the "Missouri Weinstrasse."

Peers is known for its general store, a railroad-era building with a room for rent upstairs. The small town of Treloar was founded in 1899 and was named for William Mitchellson Treloar, a professor of music at Hardin College and a U.S. Representative from Missouri. Berger is a small city first settled in 1818. It is home to the Bias Vineyards and Winery and is part of the Herman American Viticultural Area.

Washington began as a boat landing on the Missouri River, and the town was officially established in 1839. John B. Busch, brother of Adolphus Busch in St. Louis, founded a brewery in Washington in 1854 and produced the first Busch Beer. Franz Schwarzer began manufacturing what would become his world-famous zither (a German, stringed musical instrument) in 1866. Henry Tibbe began making corncob pipes in 1869, and Washington became known as the "Corncob Pipe Capital of the World." Today Washington holds the state record for the most buildings on the National Register of Historic Places.

Miller's Landing was founded in 1836 as a riverboat stop. Phillip Miller, the town's founder, operated a wood yard on the river to fuel the steamboat trade. Explorer John Colter, the first white man to explore the Yellowstone area, settled, died, and is buried nearby. German immigrants, many of them from around Borgholzhausen, Germany, also helped settle the area. In 1856, the town's name was changed to New Haven. Borgholzhause maintains ties to New Haven as the town's sister city.

Today, Hermann hosts many annual events, including Maifest, Oktoberfest, and Wurstfest, a celebration of the art of German sausage making. Approximately a quarter of a million people visit Hermann each year to enjoy a taste of "Little Germany" in the heart of Missouri wine country.

Although many tourists come to this area to visit the wineries and vineyards, at least as many come to experience the history and sample the German culture and traditions still thriving in many of these small towns along the way. The beautiful vistas, temperate spring and summer climate, and cheerful residents whose virtues Gottfried Duden once extolled still abound and are just a few of the many delightful aspects of Missouri Wine Country from St. Charles to Hermann.

One

St. Charles, Defiance, Matson, Augusta

By the 1860s, grape vineyards were plentiful in Augusta. The Augusta Wine Company was founded in 1867, and the following year, the company built the Augusta Wine Hall across from the town's public square. By 1876, the Augusta Wine Company shipped 20,000 gallons of wine out of the barrels in the two cellars beneath the hall. Political meetings, elections, and social gatherings were also held on the first floor of the wine hall. The building is pictured here around 1890 in a photograph by Ralph Goebel. (Courtesy of Anita M. Mallinckrodt, Friends of Historic Augusta.)

Henry Jaeger's home in St. Charles County is being moved by a three-horse team in this late-1800s photograph. (Courtesy of the Dora Utlaut Schneider Collection, St. Charles County Historical Society.)

Adam Loeffert and his wife pose for a photograph in the late 1800s in front of what appears to be a grape vineyard. (Courtesy of the Dora Utlaut Schneider Collection, St. Charles County Historical Society.)

A train heads for the station in Matson, Missouri, in this Ralph Goebel photograph. (Courtesy of Anita M. Mallinckrodt, Friends of Historic Augusta.)

The Heller Hotel was located in this Augusta building around 1900. (Courtesy of St. Charles County Historical Society.)

The streets of Augusta were unpaved in the early 1900s in this Ralph Goebel photograph. (Courtesy of Anita M. Mallinckrodt, Friends of Historic Augusta.)

In the early 1900s, T. R. Sand Works, in Klondike, Missouri (between Defiance and Augusta), was a silica sand quarry. In 2004, a 250-acre park was opened at the site, which includes nature trails, picnic areas, camping sites, a fishing lake, and a boat ramp. Ralph Goebel took this photograph. (Courtesy of Anita M. Mallinckrodt, Friends of Historic Augusta.)

A German Day Parade takes place in Augusta in the early 1900s in this Ralph Goebel photograph. (Courtesy of Anita M. Mallinckrodt, Friends of Historic Augusta.)

On November 9, 1909, the Christ Lutheran Church in Augusta installed a new modern stove. Ralph Goebel was the photographer. (Courtesy of Anita M. Mallinckrodt, Friends of Historic Augusta.)

The Matson railroad station can be seen in this Ralph Goebel photograph. Matson is an unincorporated community about 3 miles south of Defiance. (Courtesy of Anita M. Mallinckrodt, Friends of Historic Augusta.)

Main Street in Hamburg is pictured here around 1910. The town of Hamburg was evacuated in 1940 and 1941, and the area was taken over by the army for the Weldon Spring Ordnance Works. The plant manufactured trinitrotoluene (TNT) and dinitrotoluene (DNT) and later processed uranium. Today the site has been decontaminated. One-and-a-half million cubic yards of hazardous waste is contained in a disposal cell, and visitors can climb stairs to the top of the 75-foot-tall mound. (Courtesy of St. Charles County Historical Society.)

Bill Zeyen, son of Nick Zeyen, is pictured here in Hamburg about 1887. (Courtesy of the Dora Utlaut Schneider Collection, St. Charles County Historical Society.)

Michael Wepprich poses with his wife and son in front of Mike's Summer Garden (later Wepprich's Wine Garden) around 1908, in this view looking north on South Main Street. The St. Charles Vintage House Restaurant and Wine Garden currently occupies the site. (Courtesy of St. Charles County Historical Society.)

Michael Wepprich holds his son, Albert, on the porch at Mike's Summer Garden around 1908. (Courtesy of St. Charles County Historical Society.)

The Kelpe Blacksmith Shop was located in Hamburg in 1910. Pictured from left to right are Ben Raus, John Mades, Jake Vogt, E. Mottert, and Ervin Bacon. (Courtesy of St. Charles County Historical Society.)

The Fritz Weyrauch family poses for a picture in front of their home in Hamburg. (Courtesy of the Dora Utlaut Schneider Collection, St. Charles County Historical Society.)

The German Evangelical church in Hamburg, pictured here around 1910, was destroyed by a fire around 1940. (Courtesy of St. Charles County Historical Society.)

Hamburg residents, from left to right, Anna, Josie, and "Grandma" Schneider pose in front of the house in 1910 where Anna and Josie were born. (Courtesy of St. Charles County Historical Society.)

Nathan Boone, son of pioneer Daniel Boone, began building this house in Defiance in 1803 on land he traded for a horse and a bridle. Daniel Boone resided here periodically and died in the house in 1820. This Ralph Goebel photograph was taken around 1915. Today the home is open to tourists. (Courtesy of St. Charles County Historical Society.)

Young sunbathers enjoy a swimming hole near the old iron bridge crossing on the road to Howell and Defiance around 1940. The photograph is from the H. A. Insinger collection. (Courtesy of St. Charles County Historical Society.)

The Wepprich Wine Garden, as it appeared in 1960 in this Ryne Stiegemeier photograph, is now the site of St. Charles Vintage House Restaurant and Wine Garden. (Courtesy of St. Charles County Historical Society.)

The grounds and hillsides of Wepprich Wine Garden can be seen in this photograph from the Martin Harding Collection taken around 1960. (Courtesy of St. Charles County Historical Society.)

Various pieces of wine-making equipment from the 1966 Augusta Winery exhibit, now housed at St. Charles County Historical Society, include a hammer used with other tools to tighten barrel hoops, an auger-tapered reamer used to drill holes in barrels, a brass wine cask spigot, and a metal corkscrew. (Authors' collection.)

The USDA Soil Conservation Service reported that the Augusta bottom road was washed out by the spring of 1973 flood. The levee running parallel to the road broke at this point. The fast moving water washed out about 400 feet of road and created a hole about 300 feet deep. A corncrib was ready to topple into the hole when the flood subsided. Les Volmert took this photograph on June 4, 1973. (Courtesy of the Tally Darby Collection, St. Charles County Historical Society.)

Another view of the Augusta levee break can be seen here on June 4, 1973. (Courtesy of St. Charles County Historical Society.)

This building was originally constructed by the Immaculate Conception Church to house teachers who taught in the New Hope School in Augusta, and later became a private residence. The photograph was taken in March 1976. (Courtesy of St. Charles County Historical Society.)

A view of the 1993 Missouri River flood is seen from the grounds of the Mount Pleasant Winery in Augusta. (Courtesy of Mary Schulenberg.)

A bottle of Concord wine from the last Wepprich vintage is on display at the St. Charles County Historical Society. The Wepprich family sold the property in 1973. A restaurant occupies the site today. (Authors' collection.)

Mount Pleasant Winery was founded in1859 by George Münch from Germany, and the original cellars were completed in 1881. Münch's son, George Muench Jr. (who anglicized his name), maintained the vineyards in the 1880s. Active in local politics, Muench was also instrumental in getting wooden sidewalks built in Augusta. Mount Pleasant closed during Prohibition and re-opened for business in 1966. The Muench wine cellar is pictured in this photograph from the 1960s. (Courtesy of St. Charles County Historical Society.)

Presses, tubs, a bottle dryer, and steamer were a few of the items bought by Mount Pleasant Wine Cellars from the estate of Alfred Nahm. The Nahm Winery was in the basement of his brick house, west of Augusta. The winery had a capacity of 3,000 gallons, and all of the work was done by hand. Nahm was a grape grower and wine maker for more than 64 years. When his vineyard was in production, his wine varieties averaged about 500 gallons to an acre. (Courtesy of St. Charles County Historical Society.)

These large casks were installed in the Mount Pleasant Wine Cellar after having been bought from the estate of Alfred Nahm. (Courtesy of St. Charles County Historical Society.)

Alfred Nahm ceased commercial production of wine sometime before 1950, although he continued to make some wine for personal use. When his vineyards were in operation, the principal grape varieties he grew were Missouri Riesling, Elvira, and Virginia Seedling. (Courtesy of St. Charles County Historical Society.)

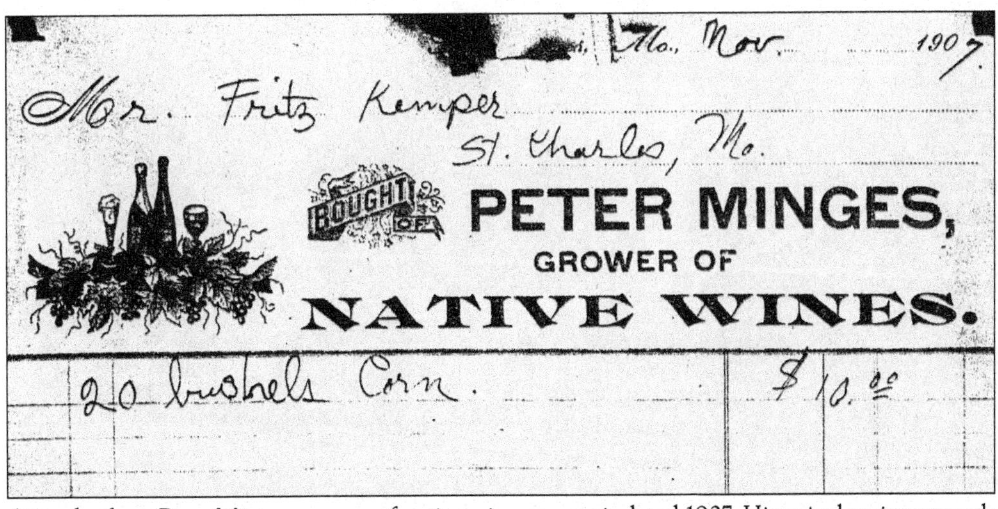

An order from Peter Minges, grower of native wine grapes, is dated 1907. Historical society records show that Minges, a German immigrant, declared his intent to become a naturalized citizen in December 1892. (Courtesy of St. Charles County Historical Society.)

The land where today stands the Little Hills Winery in St. Charles was reportedly purchased in 1805 for $1,500, with consideration of $500 in animal skins. The building once housed a pharmacy and later a meat locker and processing plant. It is said that bootleg whiskey was sold through the back door during Prohibition. Today the site is home to a restaurant and winery. The vineyard is in Eolia, Missouri. (Authors' collection.)

St. Charles Vintage House Restaurant and Wine Garden is located on Main Street, at the site of what was once Wepprich's Wine Garden. The Wepprich family sold the winery in 1973. Although wine is sold at the restaurant today, it is no longer made there. (Authors' collection.)

The site of the Yellow Farmhouse Vineyard and Winery in Defiance was originally a blacksmith shop and then an old general store. In the late 1800s, horsemen would use it as a stopping off point before heading into St. Louis. The foundation of the general store still lies just below the surface of the front lawn. The house and the barn, which is the winery building, were constructed in the 1920s. (Authors' collection.)

In the 1870s, a freed slave named Joseph Chandler fled the south after the Civil War and settled near the present town of Defiance. He befriended a nearby family and worked on their farm for several years. They later deeded him 40 acres of their land. Chandler died in 1952 at the age of 103. The Chandler Hill Winery was built on the land he once owned, and some of his possessions are on display at the winery. (Authors' collection.)

Sugar Creek Winery was founded in 1994 in Defiance. The owners moved from Kirkwood, Missouri, to Defiance to open the winery. (Authors' collection.)

Augusta Winery was established in 1988 by Tony Kooyumjian on the bluffs overlooking the Missouri River Valley. (Authors' collection.)

Mount Pleasant Winery, founded in 1859 by George Münch, a German immigrant, is still in operation today in Augusta. (Authors' collection.)

Montelle Winery was founded in 1970 and is part of the Augusta wine district. Montelle is also the first winery in the state to open a distillery. Four different types of brandy are produced on site. (Authors' collection.)

Louis P. Balducci was involved in the Missouri wine industry from 1946 until 1979. He was highly respected as a wine specialist in the area. The Louis P. Balducci Winery is located in Augusta. (Authors' collection.)

Two

Dutzow, Marthasville, Peers, Treloar

Harvey Griswold built his home in Marthasville around 1843, according to a carved stone on the house. William Ramsey Sr. acquired the land in 1816, and it was later deeded to John Young, who laid out the plats for Marthasville. John and his wife, Martha, deeded the property to Harvey Griswold. Originally there were slave quarters built in the back of the house. (Courtesy of Warren County Historical Society.)

A brick building is featured in a photograph of Frederick Muench's property near Dutzow. Friedrich Münch (later anglicized to Frederick Muench) was a follower of Gottfried Duden and was one of the first settlers in Dutzow. He was co-leader of the 1834 Giessen Society and a well-known writer and lecturer. (Courtesy of Warren County Historical Society.)

Tilman Cullom Sr. built his home in the Gore area west of Marthasville in 1831 after emigrating from Kentucky. Prior to coming to Missouri, Cullom had been a city attorney, Illinois state representative, two-term Illinois governor, and U.S. senator. After settling in his new home, he became a farmer and a businessman; he was named president of the Warren County Court upon its organization in 1833. The Tilman Cullom house was torn down in 1910, and a new house was built on the same site using some of the original walnut timber of the first home. It is assumed the people pictured are Cullom and his family. (Courtesy of Warren County Historical Society.)

The Krueger Bluff Farm in Charrette Township near Marthasville was purchased from the government in 1844 by Conrad Mische. An interesting geological structure on the land is a canyon created by a stream flowing out of the hills, running between two bluffs about 30 feet apart. Because of this formation, the property has also come to be known as "Canyon Farm." (Courtesy of Warren County Historical Society.)

Schaaf's Mill in Marthasville, later called the Schaaf-Bierbaum Mill, was started by C. H. Schaaf. A German emigrant, Schaaf opened the first horse mill in Warren County in 1841. He changed the mill to an ox mill in 1846. In 1854, it was replaced by the first steam mill in the county. Herman H. Bierbaum came to the area from Germany in 1845 and learned the milling business from Schaaf. He married Schaaf's daughter Anna in 1848. Bierbaum bought and operated the mill on the hillside for many years. His son Carl operated the mill from about 1905 until 1913. In all, the mill had been in operation for about 72 years when it was shut down. (Courtesy of Warren County Historical Society.)

Posing in front of a home built by Otto Ahmann in 1875 are (from left to right) Ida Mittler, Otto Ahmann, Clara Muench, ? Ahmann, Elise (Hildebrand), and ? Lienehe. (Courtesy of Warren County Historical Society.)

With his son-in-law Flanders Callaway and a few friends, Col. Daniel Boone established Callaway Post near La Charrette Village between 1795 and 1799. Callway Post was possibly the westernmost settlement of Americans at that time. The post was a group of cabins, arranged to provide some security from Indian attacks. The two-story log house was built in 1812. Daniel Boone lived with the Callaways in the large house after his wife, Rebecca, died in 1813. (He also occasionally resided with his son Nathan at his home in Defiance, where he died in 1820.) The Callaway home was located a mile south of Marthasville and was demolished in September 1971. (Courtesy of Warren County Historical Society.)

A threshing crew and hay makers work in a Treloar field in the late 1800s. (Courtesy of Warren County Historical Society.)

Luppold Store in Case was built in 1893 by Matthias Luppold around the time the Katy Railroad was constructed in the southern part of Warren County. Luppold had emigrated from Germany in the late 1850s and moved to Budd in the late 1860s, where he continued his trade as a blacksmith. Budd was a boat landing on the river between Case and Gore, about 3 miles east of the location of this store at Case. The Luppold family operated the store until 1947. The building is located on Highway 94 about 4 miles east of Hermann. (Courtesy of Warren County Historical Society.)

Teachers and students pose outside Gore School, at the intersection of what is now Highway 94 and Pea Ridge (Christmas Tree) Farm Road, just north of what was once Budd, Missouri. The small town disappeared when the Katy (Missouri-Kansas-Texas) Railroad was built in the late 1800s. Gore is located across the Missouri River from Berger. (Courtesy of Warren County Historical Society.)

Members of the Farmers Institute in Marthasville are pictured in 1895. (Courtesy of Warren County Historical Society.)

This aerial view of downtown Marthasville, first published in the *Marthasville Record*, was taken in either 1885 or 1895 by Ralph Goebel. (Courtesy of Warren County Historical Society.)

The Rekate Store in Pinckney, northwest of Treloar, was probably built about 1871 and was originally owned and operated by Anton Rekate. Farmers could sell their produce and wares at the store, as well as purchase work clothing, groceries, and other items. For years, Rekate Store was considered an important rural store in the county. Anton's son Herman took over in 1890 when his father died. An addition to the building was constructed about 1900. (Courtesy of Warren County Historical Society.)

Members of the Dutzow Cornet Band pose for a group photograph that appears to have been taken in the late 1800s. From left to right, band members are (first row) Ed Feldmann and Gus Berg; (second row) Fritz Dickhaus, Ben Nagel, August Riefenberg, and Charles Rexroth; (third row) Professor Husles, Anton Diermann, Otto Hinnah, Henry Rohe, George Schultz, John Schopp, and Alvin Dickhaus. (Courtesy of Warren County Historical Society.)

The Farmers Institute was held in Morhaus Hall in Marthasville. This Ralph Goebel photograph was taken around 1895. (Courtesy of Warren County Historical Society.)

Ralph Goebel snapped this photograph at the Marthasville Depot around 1900. Marthasville was named for the wife of the town's father, John Young. (Courtesy of Warren County Historical Society.)

Cora Schneider poses at the Marthasville Depot. Marthasville is about 45 miles southwest of St. Charles. This depot was once part of the Missouri, Kansas, and Texas (MKT) Railroad until the line was abandoned in 1986. Today it is part of the Katy Trail State Park. (Courtesy of Warren County Historical Society.)

This picture of the railroad depot at Case during the late 1800s or early 1900s features, from left to right, ? Haberthier, Hugo Weber, Mike Split, Cliff Watson, and Bill Haselroth. Case is a small town west of Treloar. (Courtesy of Warren County Historical Society.)

The original bank, store, and post office in Treloar, Missouri, are seen here as they appeared in the early 1900s. Treloar was founded in 1899. (Courtesy of Warren County Historical Society.)

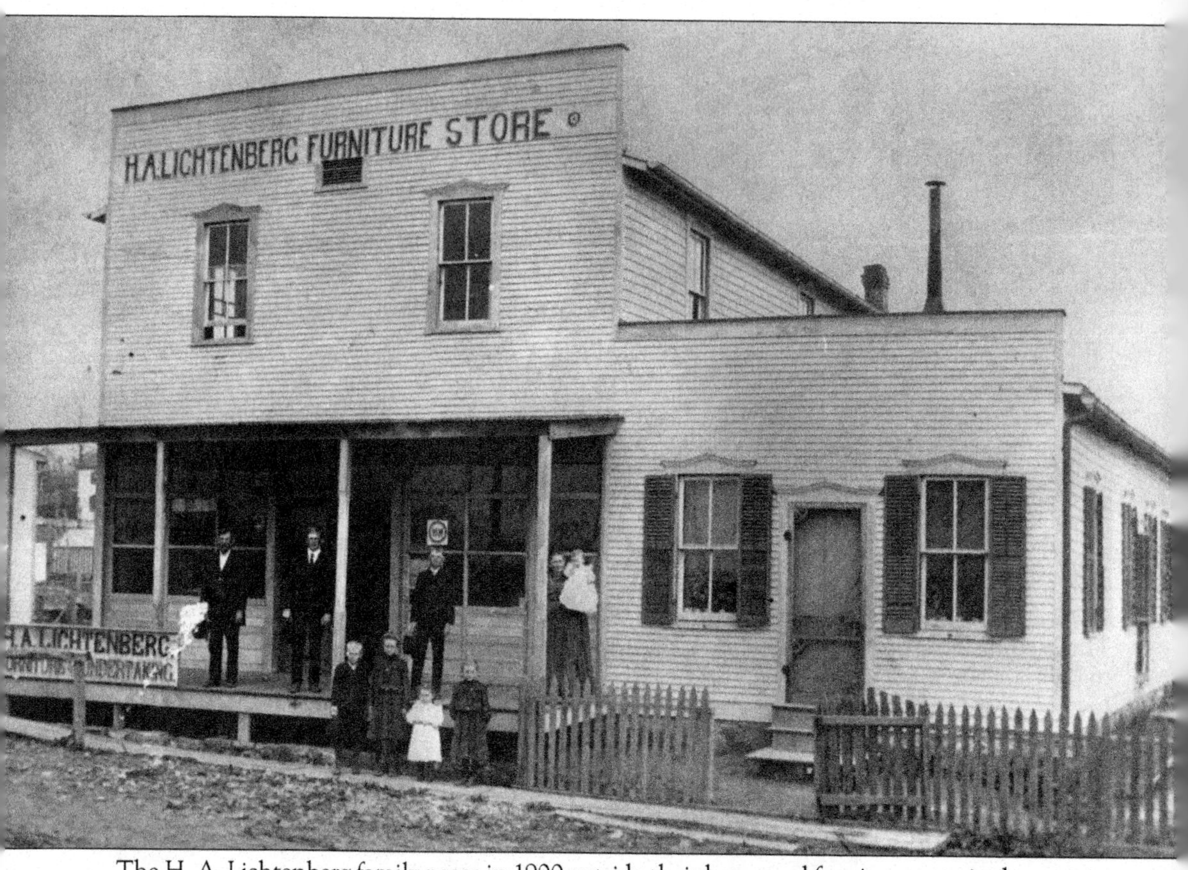

The H. A. Lichtenberg family poses in 1900 outside their home and furniture store in downtown Marthasville. The furniture and undertaking establishment was located on the corner of Two and Depot Streets. Pictured, from left to right, are (first row) Edwin Lichtenberg, Laura (Zertenna), Mildred (Copeland), Stella (Johnson); (on porch) H. A. Lichtenberg, Fred W. Lichtenberg, Gus R. Lichtenberg, Louise (Alberswerth) Lichtenberg, and Clarie (Ahmann) Lichtenberg. This photograph was originally published in the *Marthasville Record*. (Courtesy of Warren County Historical Society.)

Marthasville Hardware, owned by Godfrey and Fritz Ahmann, shared a building with the first bank in Marthasville in 1902. Godfrey Ahmann is on the far right. (Courtesy of Warren County Historical Society.)

Students and their teacher at the Marthasville public school pose for a photograph around 1900. (Courtesy of Warren County Historical Society.)

Pictured in front of the Gore Store in 1905 are, from left to right, Mrs. Eisemann, Mr. Eisemann, Paul Reinhart, Ed Reinhart, Emmett Reinhart, and Mrs. Reinhart. (Courtesy of Warren County Historical Society.)

Peers General Store has been a popular shopping spot and gathering place for many years and is still in operation today. Travelers along the Katy Trail can rent the upstairs room for a night. (Courtesy of Warren County Historical Society.)

Kites Bridge over Charrette Creek is pictured here in 1904. (Courtesy of Warren County Historical Society.)

In 1913, Evangelical and Reformed Church in Marthasville (now St. Paul's United Church of Christ) celebrated its 50th anniversary. Pictured from left to right are (first row) Julius Lagemann, John Rettke, William Riske, Pastor Theo H. Hoefer, William Dickmann, William Ahmann, and William Ottermann; (second row) F. G. Ahmann, C. Stegen, William Hinnah, L. Foeller, Frank Riemeier, Otto Ahmann, Henry Eichmeyer, William LaGemann, and Henry (H. W.) Bierbaum. (Courtesy of Warren County Historical Society.)

The Daniel Boone monument in the Bryan Farm Cemetery in Marthasville is believed by many to be the final resting place of pioneer Daniel Boone. Boone came to the area in 1795 or 1799 and established Callaway Post near La Charrette Village. It is said that upon his wife's death in 1813, he chose a tree-covered knoll by Tuque Creek on the Bryan farm for her resting place. He was buried beside her when he died in 1820. His remains (or some of his remains) may have been moved to Kentucky about 25 years later, but it is believed that some or most of his body is still buried at this Missouri cemetery. The original plaque was stolen in 2008 but has since been replaced. (Courtesy of Warren County Historical Society.)

The dedication of the Daniel Boone monument in 1915 in Marthasville attracted many visitors. A row of cars is parked nearby as the dedication gets underway. The monument was erected by the Daughters of the American Revolution. (Courtesy of Warren County Historical Society.)

Young women participate in a Maypole dance in Marthasville in May 1932. (Courtesy of Warren County Historical Society.)

A group of Marthasville students gathers for a photograph in front of the old frame school around 1932. (Courtesy of Warren County Historical Society.)

Boys from Charlotte West's 1932 class in Marthasville included, from left to right, (first row) Norman Lichtenberg, John Edward McVey, Elmer Mutert, Wilfred Ahmann, and Leroy Brakemeyer; (second row) Leroy Eichmeyer, Wilbert Wegener, Delmont Lichtenberg, Hadley Backhaus, Billy Allen Berg, Lester Koch, Forest Roloff, and Ralph Schoppenhorst. (Courtesy of Warren County Historical Society.)

Members of St. Anthony's Church in Case, Missouri, gather outside the church as it appeared around the 1940s. Case, Missouri is west of Treloar, on the north side of the Missouri River not far from Hermann. (Courtesy of Warren County Historical Society.)

The sun sets on a late fall day at Thierbach Orchards and Berry Farm in Marthasville. This path leads through the peach orchard. (Courtesy of Bob and Judy Gross.)

An original 1860s log cabin, owned by Bob and Judy Gross, is still in use as a bed and breakfast in Marthasville. (Courtesy of Bob and Judy Gross, Critter Cottage.)

Blumenhof Winery is in Dutzow, the oldest German settlement in Missouri. *Blumenhof* translates from German and means "court of flowers." Blumenhof Winery was founded in 1979, and its first vintage was produced in 1986. (Authors' collection.)

Three

Washington

Anton Jasper was founder of Jasper and Sons in the 1850s. In 1897, his sons, George F. and H. A. Jasper, took over. The men were blacksmiths and sold agricultural equipment. It is believed there was an elevator in this building that could raise tractors up to the second floor. (Courtesy of Washington Historical Society.)

John Baptiste Busch, older brother of St. Louis' Aldophus Busch, founded Washington's Busch Brewery in 1854. Low temperatures in bricked basement cellars were used to age lager beer. During Prohibition, the brewery made a beer substitute and was the place to go for a block of ice. However, the brewery never recovered from Prohibition and closed in 1953. (Courtesy of Washington Historical Society.)

Pictured is a Washington Lager Beer label from the 1890s. (Courtesy of Washington Historical Society.)

The engine room of Busch Brewery in Washington is pictured here when it was in operation. (Courtesy of Washington Historical Society.)

John B. Busch was the founder of Washington's Busch Brewery. (Courtesy of Washington Historical Society.)

Ben Jasper and Son was a business in the western part of Washington during the late 1800s. The Odd Fellows Cemetery can be seen in the background. This area was dubbed "Jasperville" by Washington locals because so many members of the Jasper family lived nearby. The road was called Pottery Road at the time of this Zoff photograph. (Courtesy of Washington Historical Society.)

The *Bright Star* was the ferryboat at Washington, Missouri, in 1863. The boat was then owned by Joseph Kettler and was one of the boats used to evacuate Washington during the Civil War. In April 1867, Joseph Kettler sold his interest in the *Bright Star* to the millers, Bleckman and Horn, of Washington. It continued in service as a ferryboat at Washington until the May Bryan took over in April 1875. The ferry's early captain was A. S. Bryan, and it was later captained by Frank Hoelscher and Robert Roehrig. This photograph was taken in the 1870s. (Courtesy of Washington Historical Society.)

The Grammar School, Washington's old public grade school, was torn down in 1957. The new high school building had just been finished, so the bricks from the old grade school were ground up to make a high school track. The grade school was built in 1879 and had a fire escape on the back that was an enclosed metal tube-like slide. It is said that the children loved to have fire drills so they could slide down the escape. This photograph was taken in the 1880s. (Courtesy of Washington Historical Society.)

Henry Tibbe lived from 1819 to 1896 and was the originator of the corncob pipe factory that would later become Missouri Meerschaum. Tibbe was a Dutchman from Enschede who immigrated to the United States when his entire village burned down. A woodworker by trade, Tibbe opened a shop on Second Street. Legend has it that a farmer brought him a corncob and asked him to whittle a pipe. (Native Americans had smoked corncob pipes for years.) Eventually, others wanted Tibbe's pipes too, and he went into business. The house still stands and faces west Fourth Street on the corner of Cedar Street. This photograph is from the estate of Walter Kahmann; the photographer was G. C. Parks. (Courtesy of Washington Historical Society.)

An unusual snowstorm occurred in May 1894, and is shown in this photograph of Main Street looking east from Elm Street. Gallenkamp Drugs, whose sign is seen above the Washington Ice buggy, burned in 1915. Schmidt Boss Jeweler is on the right, and the building still stands. The building to the left of Gallenkamp's was torn down to make a bank. (Courtesy of Washington Historical Society.)

Looking east on Second Street from Lafayette Street, one can see the Central Market on the right. E. A. Fricke was the proprietor. Children called it the "Cow Head Store" because it had an artificial cow head fastened to the front. The building on the left next to the Central Market was built in 1852. Electricity came to Washington in 1893, so this photograph was probably taken around 1900. (Courtesy of Washington Historical Society.)

Deutsche Sage (German Tradition) is stenciled on the sign above the girls in this 1906 German Day parade float. This view is in Old City Park at the end of Second Street. (Courtesy of Washington Historical Society.)

Before it was called International Shoe Company, this operation was named Roberts, Johnson, and Rand Shoe Company. The factory was built in 1907, and this photograph was probably taken shortly after that. The complex of buildings still exists at West Second Street. During World War II, the shoe company manufactured army boots for the military and was the largest of the Washington shoe factories. International Shoe Company went out of business in 1960. (Courtesy of Washington Historical Society.)

Anton Arnold Tibbe stands with his hand on the doorpost in front of the corncob pipe company. The structure was built by his father, Henry Tibbe, at Front and Cedar Streets in 1883. This photograph was taken after 1907, when H. Tibbe and Son became Missouri Meerschaum. Meerschaum is a Turkish clay used in high-grade pipes. The word means "sea foam" in German. Tibbe felt his light, porous pipe design and its cool smoke was similar to the meerschaum pipes, so he named his pipes Missouri Meerschaums. G. C. Parks was the photographer. (Courtesy of Washington Historical Society.)

The Kahmann building was constructed at Second and Elm Streets in the late 1890s and was called Kahmann's Store when it was built. This photograph was taken sometime after 1922 when the building became J. C. Penney's first location in Washington. (Courtesy of Washington Historical Society.)

City Hall at Jefferson and Fourth Streets is pictured here in the mid to late 1920s. The original 1850s building at this site was replaced with this one, which is still standing. An addition was added in 1923. (Courtesy of the Washington Historical Society.)

This Bank of Washington photograph was snapped by Edward Zoff. The bank appears as it was constructed in 1923 at the southwest corner of Main and Oak. The building style was typical for banks at that time. The building still exists, but a new five-story bank building is constructed around it. Some parts of old bank may be visible inside the new building. (Courtesy of Washington Historical Society.)

The Roberts, Johnson, and Rand Shoe Company, shown here around 1930, employed close to 1,600 employees during World War II. It was later renamed International Shoe Company. (Courtesy of Washington Historical Society.)

Santa Claus stands beside a biplane that flew him into town in 1932. Kruel's Store arranged the transportation. (Courtesy of Washington Historical Society.)

Washington photographer Ed Zoff spends Christmas with his granddaughters, Susan and Judy, sometime during the early 1930s. Edward Zoff died in 1936. (Courtesy of Washington Historical Society.)

The windows at Otto and Company in the early 1930s display Gold Seal Congolium Art Rugs. Otto's was located at 127 Elm Street. (Courtesy of Washington Historical Society.)

The Franklin County Bank celebrated its centennial in November 2009. This Zoff photograph was taken during the early 1930s. (Courtesy of Washington Historical Society.)

Nieburg and Vitt, Inc. Furniture Store, shown here in the early 1930s, also had a funeral home up the street. The building was constructed in 1926. Furniture and funeral homes often were co-owned. Nieburg and Vitt Funeral Home is still in existence, but the furniture store is not. (Courtesy of Washington Historical Society.)

Spectators watch the American Legion Parade in Washington in 1931. This view is looking north on Elm Street from Second Street; the Missouri River is straight ahead. The train depot can be seen toward the end of the street. (Courtesy of Washington Historical Society.)

In this Edward Zoff photograph of the 1931 American Legion Parade, Otto and Company Furniture Store can be seen in its fourth location. The business was established in 1845 and occupied this building from the 1890s until about 2004 or 2005. Otto is retired, and the building is currently used for office space and apartments. The building on the right was built by Anton Kahmann around 1868 and is also still standing. (Courtesy of Washington Historical Society.)

The Sinclair filling station at the northeast corner of Fifth and Jefferson Streets may have been owned at this time by George Hausmann. This Zoff photograph was taken about 1931. The location is now a tire company. The garage portion of the structure is still standing, although it has been renovated. (Courtesy of Washington Historical Society.)

The C. A. Krumsick/Dodge Brothers Motorcar Company was one of the very first Dodge agencies and is shown in this Zoff photograph as it appeared in 1931. A new Plymouth is advertised in the window and cars are being delivered for sale. Immanuel Lutheran Church is seen on the right, and the old elementary building rises above the auto building in back. (Courtesy of Washington Historical Society.)

The sign on this residence states that it is a fireproof model home, erected under tri-ply construction. John M. Schaper was the architect and built it in 1931 for his brother, Randolph, a county judge. The house is still standing. (Courtesy of Washington Historical Society.)

Construction of the Washington City Auditorium began in 1935. This project was sponsored by the WPA (Works Progress Administration) under Pres. Franklin Delano Roosevelt. The unusual structure included a lamella roof, a vaulted roof of crisscrossing beams. The building still exists in Old City Park at Second and High Streets. It is occasionally used for the annual Christmas banquet for the city employees and is also available for rent. The photograph was taken by Edward Zoff. (Courtesy of Washington Historical Society.)

The Kruel's Store building on Main Street, which dates to 1877, still exists. Kruel's was in operation from the early 1900s until the late 1960s or 1970s. Fred Kruel was the owner when this picture was taken during the 1930s. The building housed the First Bank of Washington in 1877. In the time between the bank's closing and the opening of Kruel's, the building housed the Voss Bazaar Store. It is currently an arts center. (Courtesy of Washington Historical Society.)

The oldest part of the Missouri Valley Creamery (on right) was built in 1916. The cornerstone from that part of the structure is now in the flower garden at the Washington Historical Society. The part of the building on the left was originally one story. The business was owned by F. W. Springer, brother-in-law to photographer Edward Zoff, and closed for business around 1960. Before then, children could stop in any time and get a free spoon of ice cream. Before the building was demolished around 2004 or 2005, a local tire company used it for storage. (Courtesy of Washington Historical Society.)

Members of the Benevolent and Protective Order of Elks pose around 1940. The building was home of the former Washington Turn Verein, a German gymnastics and athletic organization that was founded in 1859 and that also sponsored dramatic productions and social gatherings. When the organization dissolved in 1932, the Elks bought the building. The Elks constructed a new structure around 1999 or 2000, and the City of Washington bought this building, demolished it, and erected a public safety building for law enforcement. (Courtesy of Washington Historical Society.)

A crowd waits for Santa Claus to arrive in front of Kruel's Store during the early 1940s. Kruel's sponsored Santa's visit every year. (Courtesy of Washington Historical Society.)

The railroad tracks in Washington, Missouri, run parallel to the Missouri River. The brick station was constructed in 1923 and was a replacement for a wooden depot that was built in 1865. The station is now served by Amtrak. (Authors' collection.)

The Washington bridge over the Missouri River was built in 1934. It is 2,561 feet long and carries one lane of automobile traffic in each direction. (Authors' collection.)

Four

NEW HAVEN AND BERGER

Front Street is busy with activity in July 1895. The historic downtown area is listed on the National Register of Historic Places. (Courtesy of New Haven Preservation Society.)

Old Wolff Mill is shown here around 1890. The lower story is constructed of stone; the upper stories are made of brick. The Wolff Mill made and shipped almost a dozen different types of flour to customers across the country. (Courtesy of New Haven Preservation Society.)

Mary and Julian Bagby pose in front of their home on Melrose Street around 1900. Their state-of-the-art buggy is equipped with fenders and a light. (Courtesy of New Haven Preservation Society.)

A driver and his horse haul a load of trees from Bagby Nursery around 1900. Founded in 1872, the Bagby Nursery grew to 500 acres and specialized in fruit trees. In 1903–1904, more than 1 million trees were shipped to 36 states. (Courtesy of New Haven Preservation Society.)

Front Street in New Haven is pictured here in the early 1900s. (Courtesy of New Haven Preservation Society.)

Wolff Milling Company, the rail yard, and the Missouri River are shown in this picture taken during the 1903 flood. (Courtesy of New Haven Preservation Society.)

Workers on *The Dauntless* steamboat carried goods off boat to higher ground near Main Street during the 1903 flood. (Courtesy of New Haven Preservation Society.)

A visitor rests against the gate at the Central Hotel in the early 1900s. (Courtesy of New Haven Preservation Society.)

A train wreck occurred in New Haven in the early 1900s. Wolff Mill can be seen in the background. (Courtesy of New Haven Preservation Society.)

A view of Front Street is seen in this 1907 photograph. New Haven was first called Miller's Landing. In 1836, Phillip Miller bought the original tract of land to use as a wood-yard to supply fuel to passing boats. New Haven was incorporated on July 12, 1881. (Courtesy of New Haven Preservation Society.)

The Berger Public School building is pictured here in 1908. Berger is northwest of New Haven, between New Haven and Hermann. (Courtesy of New Haven Preservation Society.)

The old New Haven public school is shown here around 1909. (Courtesy of New Haven Preservation Society.)

A 270-foot-long suspension bridge was built around 1900 and taken down around 1930. At that time, a deep ravine cut across town, and schoolchildren used the bridge to get to and from school. The bridge was later replaced by a shorter fixed wooden bridge until Bates Street was extended. This photograph was taken around 1910. (Courtesy of New Haven Preservation Society.)

A pedestrian walks across the suspension bridge. (Courtesy of New Haven Preservation Society.)

From left to right, Myrtle Wolf, Robert Gruebel, and Hazel McKeehn participate in a school play in 1912. (Courtesy of New Haven Preservation Society.)

To stabilize the bank in New Haven, workers added rock to the riverbank to halt erosion. Before that, the area had a large sandy beach with giant cottonwood trees. Children liked to play ball on the beach. In 1913, the river current changed and began washing away the shoreline. The giant trees were washed away, and the buildings on Main Street were endangered. The New Haven Flood Control Levee was completed by the U.S. Army Corps of Engineers in April 1955 and runs the entire length of downtown New Haven. It was the only levee in the area to hold during the flood of 1993. (Courtesy of New Haven Preservation Society.)

Willow branches were woven into large mats and sunk with tons of rock to stabilize the riverbank. The job was started in 1913 and finished in 1915 when this photograph was taken. (Courtesy of New Haven Preservation Society.)

The flood of 1915 is shown in this photograph. (Courtesy of New Haven Preservation Society.)

In 1916, the Central Hotel can be seen on the right, and a factory is on the left. The Central Hotel was one of the first hotels in town, across from the railroad and the downtown area. It is currently a bed and breakfast. (Courtesy of New Haven Preservation Society.)

Students plant a school garden in 1917. Today the agricultural industry makes use of the fertile soil in the Missouri River area. (Courtesy of New Haven Preservation Society.)

A group of women gathers on the riverfront in May 1917. The riverfront today boasts a picturesque downtown, museums, shops, and a levee walk along the river. (Courtesy of New Haven Preservation Society.)

A float travels with the New Haven parade on February 22, 1917. (Courtesy of New Haven Preservation Society.)

A New Haven High School dance class practices on May 18, 1917. (Courtesy of New Haven Preservation Society.)

Teacher Mary Diggs poses with her first- and second-grade classes in the fall of 1917. Some of the children's hats may have been made in the Langenberg factory. (Courtesy of New Haven Preservation Society.)

A ground-breaking ceremony is held for the addition to St. Peters United Church of Christ. The date is unidentified. (Courtesy of New Haven Preservation Society.)

The New Haven High School basketball championship team photograph was probably taken between 1912 and 1920. The team members are, in no particular order, Walter Diggs, Mose Bagby, Harold Bagby, Paul Altheise, and Herbert Witinson. The coaches are, in no particular order, Mr. Settle and Mr. Livesey. (Courtesy of New Haven Preservation Society.)

The New Haven Public School was open from 1883 until 1985. Originally scheduled to be demolished and used for bricks, the building was saved and is now home to the New Haven Preservation Society. This photograph was probably taken between 1900 and 1920. (Courtesy of New Haven Preservation Society.)

A ferryboat stops at New Haven in 1920. Pictured are, from left to right, Fred Schroer, unidentified, Bill Gerding, Fritzie Koch, and Lillian Chroer. (Courtesy of New Haven Preservation Society.)

The *Tilda Clara* was literally a horse-powered boat. Horses were tethered to a large wheel that turned as they walked in circles, which turned the paddle wheel, which generated power. New Haven can be seen on the other side of the river. Pictured on the boat are, from left to right, Capt. Chrys Schwentker, Eddie Schwentker, Mr. Schwert, Mr. Haase, and Mrs. Utlaut. (Courtesy of New Haven Preservation Society.)

A recreational football squad poses for a team picture in New Haven during the fall 1922 season. (Courtesy of New Haven Preservation Society.)

Langenberg Hat Company began in 1928 and employed 150 workers by 1956. The company imported rabbit skins and removed and processed the fur. At its peak, the factory produced almost half-a-million hats per year. Langenberg Hats were sold throughout the United States and the world. The company is pictured here around 1940. (Courtesy of New Haven Preservation Society.)

Highway 185 was an unpaved road in New Haven in February 1928. New Haven today is a town with big-city facilities and a small-town way of life. (Courtesy of New Haven Preservation Society.)

A train wreck occurred in downtown New Haven in the 1930s. (Courtesy of New Haven Preservation Society.)

Another train wreck occurred in 1946. The train, which was carrying hundreds of onions and blue plums, derailed and took out part of the train station. Townspeople gathered the spilled plums and ate them for days. The onions caused quite a smell and made the townspeople's eyes water, so eventually the city hauled them away to a dump. (Courtesy of New Haven Preservation Society.)

The Hoelscher-Schwentker Service Station appears here in the 1940s. It is still a service station but no longer sells gasoline. (Courtesy of New Haven Preservation Society.)

A fire started in the C. J. Harris Lumber Company in June 1950 and spread to a neighboring grain elevator, machine shop, and hotel. Water was pumped from the Missouri River to supplement city water supply. Fire trucks from five communities fought the blaze. (Courtesy of New Haven Preservation Society.)

Spectators watch as the fire departments fight the downtown blaze in 1950. (Courtesy of New Haven Preservation Society.)

This 1950s photograph shows the northern portion of Wolff Flour Mill, which shipped almost a dozen varieties of flour over most of the country. Grain was ferried across the river from Treloar and up from southern Franklin County. The family who originally owned the mill is no longer in town. The plant was torn down in 2009 due to disrepair. (Courtesy of New Haven Preservation Society.)

A funeral buggy joins the Centennial Parade in 1957. (Courtesy of New Haven Preservation Society.)

John Colter was among the first recruits of Lewis and Clark, partly because of his hunting ability. Colter is credited as the first white man to have discovered Yellowstone, an area that later became America's first national park. His accounts of bubbling hot springs would lead writers of the era to dub the discovery "Colter's Hell." The explorer eventually settled along the Missouri River near present-day New Haven, Missouri, where he died in 1812 after spending several months serving under the command of Nathan Boone, son of pioneer Daniel Boone. Today a boulder from the Yellowstone River region, donated by the people of Montana and shipped to New Haven in 2003, serves as a memorial to John Colter, Missouri's famous mountain man. (Authors' collection.)

Some of the original buildings on Front Street in New Haven are still standing today. Many structures in the downtown area date from the late 1800s. (Authors' collection.)

Bias Winery in Berger is a family-owned operation. The wines are made from the family's own vineyards at the same site. Bias was the first winery in Missouri and the second in the country to operate as a combined winery and microbrewery. (Authors' collection.)

The Bommarito Estate Winery is in New Haven. The Bommarito family emigrated from Sicily in 1902, and food and wine-making have always been a family tradition. The winery was started in 1994 with the purchase of 21 acres in the Missouri wine district. (Authors' collection.)

Röbller Winery and Vineyard near New Haven produced its first vintage in 1990. The first vineyards were planted in 1988. It is also family-owned. (Authors' collection.)

Five

Hermann

Michael Poeschel was founder of Stone Hill Winery in 1847. (Courtesy of Stone Hill Winery.)

Otto Gula was an early manager of Stone Hill Winery. He is dressed in a uniform worn while a student at Heidelberg University in Germany. (Courtesy of Historic Hermann, Inc.'s Museum at the German School.)

Employees of Stone Hill Wine Company pose in 1890. Pictured from left to right are (first row) Sam Baumgaertner Jr., Gustave A. Freund, Fred Fluhr, Carl Sauer, and Oswald Fluhr; (second row) Casper Leibach, Sam Baumgaertner Sr., Albert Drusch, Hermann Drusch, Henry Drusch, John Huck, and Gottlieb Vonar; (third row) Anton Hagen, Otto Drusch, William Fritzemeyer, Charles Strecker, Louis Kolb, and Anton Gebhardt. (Courtesy of Historic Hermann, Inc.'s Museum at the German School.)

This Mountain Pearl Wine label is from Stone Hill Winery's early days. Stone Hill was founded in 1847 and grew to be the second largest winery in the country. (Courtesy of Stone Hill Winery.)

A horse and buggy travels down Fourth Street in Hermann in 1890. The German School is on the right. St. George Catholic School is straight ahead. (Courtesy of Gasconade County Historical Society.)

A crowd gathers at an early wine tasting in Hermann during the 1800s. (Courtesy of Historic Hermann, Inc.'s Museum at the German School.)

George Stark arrived from Germany in 1867. As president of Stone Hill Winery, Stark helped to grow the business into one of the largest wineries in the world. (Courtesy of Historic Hermann, Inc.'s Museum at the German School.)

The Hermann Apostel Band was formed in 1885. ("Apostel" is the German spelling.) All of the band members' instruments were purchased in Germany and shipped to Hermann. There were 12 musicians in the band, 1 for each of the 12 apostles. (Courtesy of Historic Hermann, Inc.'s Museum at the German School.)

Pictured is an early label from Sohns' Winery. Henry Sohns was a native of Baden who came to Hermann in 1866. His wines were made from Virginia Seedling, Ives Seedling, and Concord and Riesling grapes. Sohns' went out of business after Prohibition. (Courtesy of Historic Hermann, Inc.'s Museum at the German School.)

This postcard scene of vineyards near Hermann from around 1900 was published by Schuster Studio. (Courtesy of Historic Hermann, Inc.'s Museum at the German School.)

In 1903, residents of Hermann opened a shoe factory. It was taken over a year later by the Peters Shoe Company of St. Louis and incorporated into the International Shoe company in 1911. It later became Florsheim. This early photograph of Hermann features the shoe factory with the courthouse in the background on the left. (Courtesy of Historic Hermann, Inc.'s Museum at the German School.)

The Hermann Bank on First Street celebrates its success in the early 1900s. The bank eventually became First Bank. (Courtesy of Gasconade County Historical Society.)

The Enterprise Military Band poses in Hermann around 1920. The band later acquired uniforms. (Courtesy of Gasconade County Historical Society.)

Friends gather for a picnic alongside the Gasconade River on June 23, 1901. Mr. and Mrs. Hans Glotle, Mr. and Mrs. E. F. Rippstein, Mr. and Mrs. John Helmers, Mary Word, and Leata Wensel Baer are among the picnickers, but their positions in the photograph are not identified. (Courtesy of Historic Hermann, Inc.'s Museum at the German School.)

Hermann City Hall, still standing today, is pictured in 1906. Plans are under consideration to make it a visitors' center. (Courtesy of Gasconade County Historical Society.)

A performing bear and his trainer pause for a drink on the streets of Hermann in the early 1900s. (Courtesy of Historic Hermann, Inc.'s Museum at the German School.)

Two men appear to be enjoying afternoon refreshments in this posed photograph from the early 1900s. (Courtesy of Historic Hermann, Inc.'s Museum at the German School.)

Today's Hermannhof Winery was once the location of Kropp Brewery. The building was constructed by Jacob Strobel in 1855 or 1856. J. M. Danzer operated this brewery in 1910 at the time of the photograph. (Courtesy of Historic Hermann, Inc.'s Museum at the German School.)

Several private wine cellars existed near Hermann before Prohibition. (Courtesy of Historic Hermann, Inc.'s Museum at the German School.)

A Sohns' Missouri Wine Company plate hangs from some early wine-making equipment. (Courtesy of Historic Hermann, Inc.'s Museum at the German School.)

The Sohns family lived in the house to the right of what was originally Sohns' Wine Cellar. This photograph was taken after Prohibition, since the signs indicate an ice company and paint distributor. (Courtesy of Historic Hermann, Inc.'s Museum at the German School.)

Mushroom farming took place in the wine cellars during Prohibition, which threatened to destroy the wine-making industry in America. The cool, dark caverns that once held wine barrels were the perfect place to grow mushrooms for commercial use. (Courtesy of Stone Hill Winery.)

The Old Timers Commercial Club poses in 1923. Pictured are, from left to right, Martin Kappelmann, O. H. Nienheuser, George Kramer, unidentified, Max Kniesche, August Habsieger, and unidentified. The two men standing in back are also unidentified. (Courtesy of Gasconade County Historical Society.)

The Hungry Five musical band is pictured here around 1932 or 1933. The band featured different musicians over the years, and there were often more than five people in the band at one time. (Courtesy of Historic Hermann, Inc.'s Museum at the German School.)

The Harmonettes, the Hermann chapter of the Society for the Preservation and Encouragement of Barber Shop Quartet Singing in America, performed under the direction of Mimi Schmidt in 1947. (Courtesy of Gasconade County Historical Society.)

Schoolchildren dance in the street in the 1950s during Maifest. Maifest began in Hermann's earliest years as a May picnic for schoolchildren. The children and teachers would parade from the German School to the park carrying American flags. There they were served pink lemonade (made pink with a touch of red wine) and knackwurst, a type of sausage. In 1952, Maifest was revived as an event for both residents and tourists and is still celebrated today. (Courtesy of Gasconade County Historical Society.)

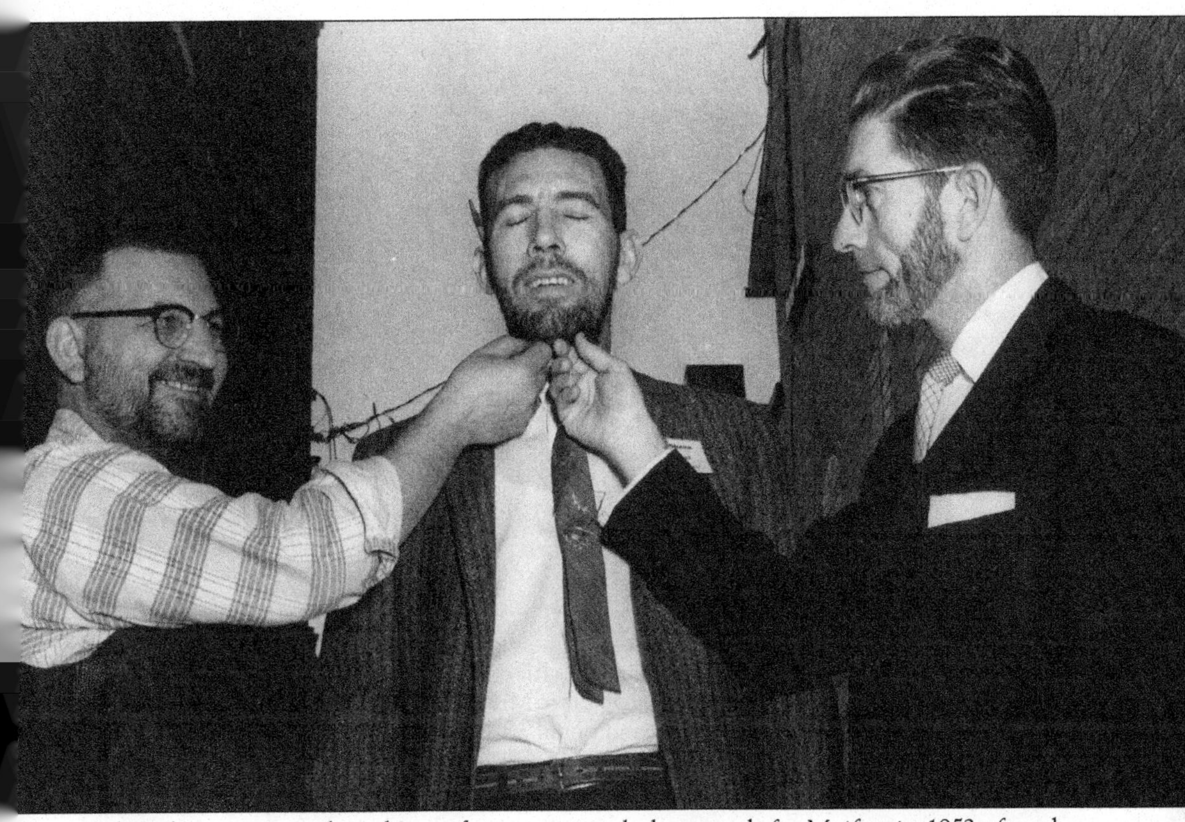

A beard contest was the subject of a promotional photograph for Maifest in 1952 after the celebration was revived as an event for residents and visitors. Pictured are, from left to right, Ray Radtke, Claude Kell, and Vic Tesson. (Courtesy of Gasconade County Historical Society.)

Sausage hangs to the right of Lucille and Paul Oelschlaeger, who are dancing at an early Maifest. Hermann is known as the sausage-making capital of Missouri. Held in Hermann each March is Wurstfest—a citywide celebration of the traditional art of German sausage making. (Courtesy of Gasconade County Historical Society.)

Anna Kemper Hesse and Clarence Hesse perform in a Maifest pageant during the 1950s. Historic pageants were a traditional part of Maifest each year from 1952 through 1964, and Anna, a key member of the preservation movement in Hermann, wrote the pageants. An artist and a teacher, Anna received the National Trust Award for Historic Preservation in 1976. Clarence Hesse was a teacher before working in the shoe factory. He also did taxidermy, oil portraits, and worked in horticulture at the Kemper Nursery. (Courtesy of Gasconade County Historical Society.)

Children perform a charming street dance at Maifest. (Courtesy of Gasconade County Historical Society.)

After Prohibition, when the making and consumption of alcohol was illegal, mushrooms began growing in wine cellars. Bill Harrison ran a steam engine that he used to kill other fungi in cellars that could affect the growth of mushrooms. The Harrison family sold the mushrooms commercially to markets beyond Hermann. The steam engine is on display here at an early Maifest at City Park. (Courtesy of Gasconade County Historical Society.)

Velten and Dorothea Rohlfing dance in the 1959 pageant, "On Flows the River." The band members, from left to right, are George Workman, Billy Sanders, Bob Kirchhofer (tuba), B. A. Wagner (clarinet), Ken Bauer, and Elmer Simon. (Courtesy of Gasconade County Historical Society.)

Jim and Betty Held, owners of Stone Hill Winery, enjoy a taste of wine during the 1960s. (Courtesy of Historic Hermann, Inc.'s Museum at the German School.)

Cooling fermenting tanks are shown in an early photograph at the Stone Hill wine cellars. Stone Hill was founded in 1847, and by the early 1900s, shipped over 1 million gallons of wine per year. (Courtesy of Stone Hill Winery.)

Jim and Betty Held, current owners of Stone Hill Winery, taste their wares around 1967. (Courtesy of Stone Hill Winery.)

Partygoers attend an event at Stone Hill Winery around the mid-1960s to early 1970s. (Courtesy of Stone Hill Winery.)

Betty Held, one of the owners of Stone Hill Winery, works in the grape vineyards in the 1960s. (Courtesy of Stone Hill Winery.)

Each one of the archways in the Apostle Cellars at Stone Hill Winery once held a large wooden vat that filled up the entire archway. On the front of each one was an image of one of the Twelve Apostles. During Prohibition, government representatives were known to tear everything apart and break up vats, so before the representatives could get to Stone Hill, its owners shipped the Apostle vats to Germany. The vats have been lost ever since. Occasionally, however, visitors to Stone Hill who have traveled to or from Germany will say they think they may have seen one or two of the vats in that country. (Authors' collection.)

Pictured here is one of the oldest wine vats that was still being used in the 1960s when the Helds bought Stone Hill. It has a small opening in the front where Betty Held crawled through to clean the inside of the vat. (Authors' collection.)

This wine press was still in use in the 1960s when Jim and Betty Held bought Stone Hill Winery and renovated the historic buildings and caverns. Stone Hill is the oldest winery in the state. (Authors' collection.)

Hermannhof Winery is one of 100 early Hermann buildings listed on the National Register of Historic Places. The building, which was a brewery and winery, was constructed in 1848 and finished in 1852. The winery was renovated in 1974. (Authors' collection.)

Adam Puchta Winery has been owned by the same family since 1855 and is the oldest family-owned winery in Missouri. Adam, along with his parents and siblings, immigrated to Hermann from Hamburg, Germany, in May 1839. The family built a log cabin on a piece of land next to the site of the current winery. (Authors' collection.)

Stone Hill Winery was established in 1847 and eventually became the second largest winery in the United States, shipping 1.25 million gallons of wine a year by 1900. During Prohibition, the dark underground cellars were used to grow mushrooms. The current owners, Jim and Betty Held, bought and renovated the property in 1965. Stone Hill is Missouri's oldest winery. (Authors' collection.)

The structure that once housed Sohns' Winery and the Sohns' family home is still standing. (Authors' collection.)

OakGlenn is situated on land once owned by George Hussmann. Hussmann was a member of the German Settlement Society of Philadelphia, which was founded to establish a German colony in America. The land was too steep and rocky for farming, so settlers planted vineyards. Hussmann's scenic property was once known as *Schou-ins-land*, which means, "look into the country." (Authors' collection.)

The town of Hermann is named for Hermann der Cherusker, a German folk hero, who led a successful battle against the Romans in 9 AD that changed the course of history. In 2009, the city of Hermann celebrated the 2,000th anniversary of Hermann's Battle of the Teutoburg Forest, and a bronze statue of the city's namesake was formally dedicated. (Authors' collection.)

Wineries and Vineyards

Augusta:
Augusta Winery
Montelle Winery
Balducci Vineyards
Mount Pleasant Winery

Defiance:
Chandler Hill
Sugar Creek Winery
Yellow Farmhouse Vineyard and Winery

Berger:
Bias Winery

Dutzow:
Blumenhof Winery

Hermann:
Adam Puchta Winery
Hermanhoff
Stone Hill Winery
OakGlenn Vineyards and Winery

New Haven:

Bommarito Estate
Röbller Winery

St. Charles:
Little Hills Winery

For more information
Hermann Wine Trail: hermannwinetrail.com
Missouri Weinstrasse: moweinstrasse.com

About the Historical Organizations

The St. Charles County Historical Society was founded in 1956 and incorporated in 1958. The organization publishes a monthly newsletter and a quarterly publication, *St. Charles County Heritage*, and sponsors various activities to promote the preservation and study of history. It is located at 101 South Main Street in St. Charles.

Warren County Historical Society is located at 102 West Walton Street in Warrenton. The museum is home to artifacts, archives, and photographs from 12 communities, including Dutzow, Missouri's oldest German settlement.

Washington Historical Society was founded in 1959. The organization has been located in the old Presbyterian Church building at Fourth and Market Streets since 1995. In addition to many exhibits and archives, the museum also houses the Four Rivers Genealogical Society.

Gasconade County Historical Society is located in downtown Hermann at 315 Schiller Street, in a 1909 brick building that was originally constructed for Farmers and Merchants Bank. The organization hosts community programs designed to make its resources, records, and artifacts readily available to the public.

Historic Hermann, Inc.'s Museum at the German School is located at Fourth Street and Schiller Street. The structure was built in 1871 and was used as an elementary school for the children of Hermann until 1955, when it was deeded to Historic Hermann, Inc. to be used for cultural and civic purposes.

New Haven Preservation Society is located on Maupin Avenue in New Haven and hosts the Springgate Museum in the old New Haven Elementary School. The school was open from 1883 until 1985, after which the preservation society saved the building from demolition.

Bibliography

Gregory, Ralph. *The German-Americans in the Washington, Missouri Area*. Washington, MO: Missourian Publishing Company, 1981.

Gregory, Ralph. *A History of Washington, Missouri*. Copyright 1991 Washington Preservation, Inc. Washington, MO.

Gregory, Ralph and Anita Mallinckrodt. *Wine-making in Duden Country*. Augusta, MO: Mallinckrodt Communications and Research, 1999.

Harrison, Samuel F. *History of Hermann, Missouri*. Hermann, MO, 1966.

Historic Sites of Warren County. Margaret C. Schowengerdt, editor. Warrenton, MO: Warren County Historical Society, 1976.

Hesse, Anna Kemper, ed. *Little Germany on the Missouri*. Columbia, MO: University of Missouri Press, 1998.

Menke, David M. *The Saga of John Colter*. New Haven, MO: Leader Publishing Co., 2003.

Schrader, Dorothy H. *Hermann Sesquicentennial*. Hermann, MO: Graf Printing Company, 1986.

CPSIA information can be obtained
at www.ICGtesting.com
Printed in the USA
BVOW07*1220160717
489421BV00012BA/102/P